This Isometric Graph Paper book belongs to:

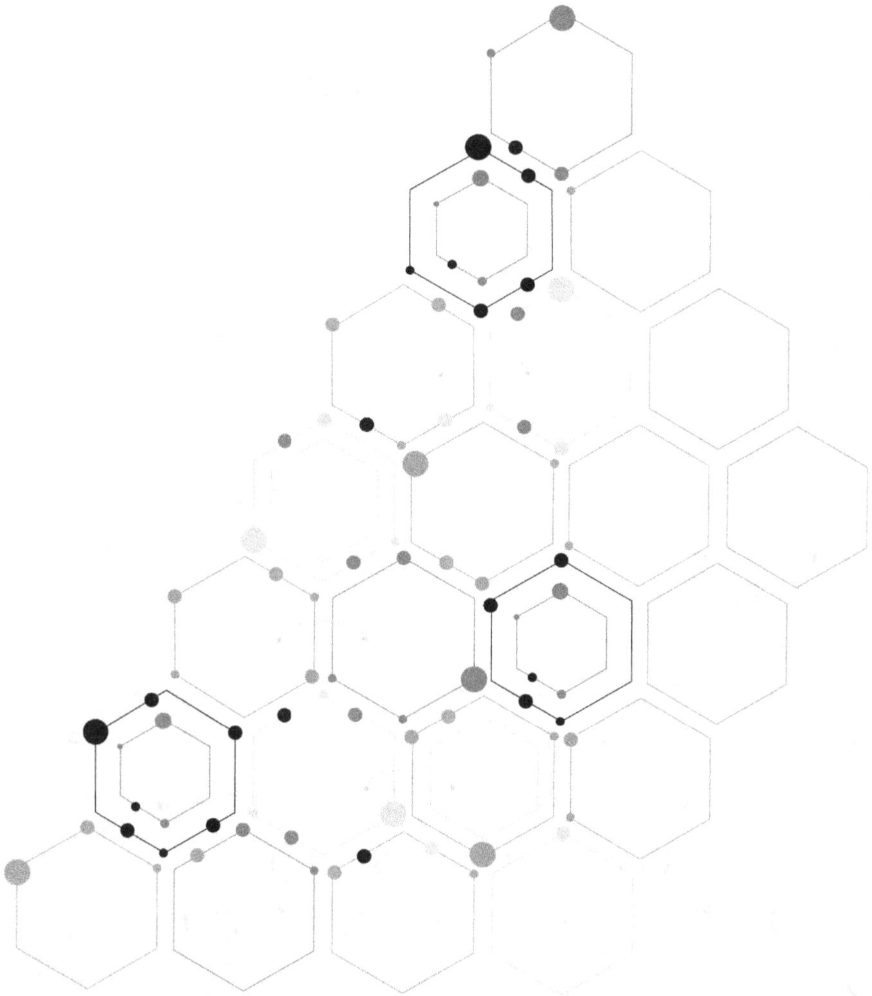

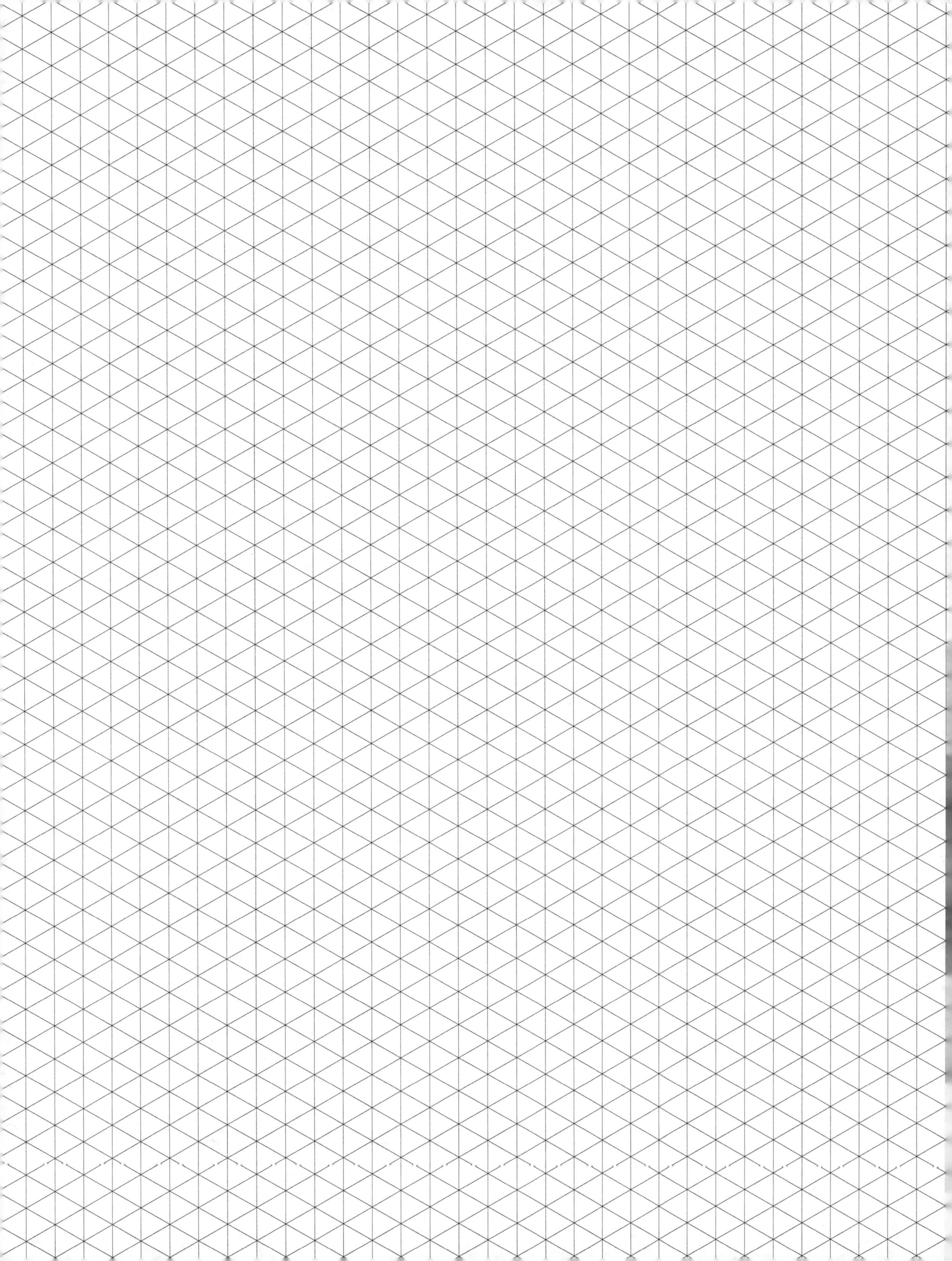

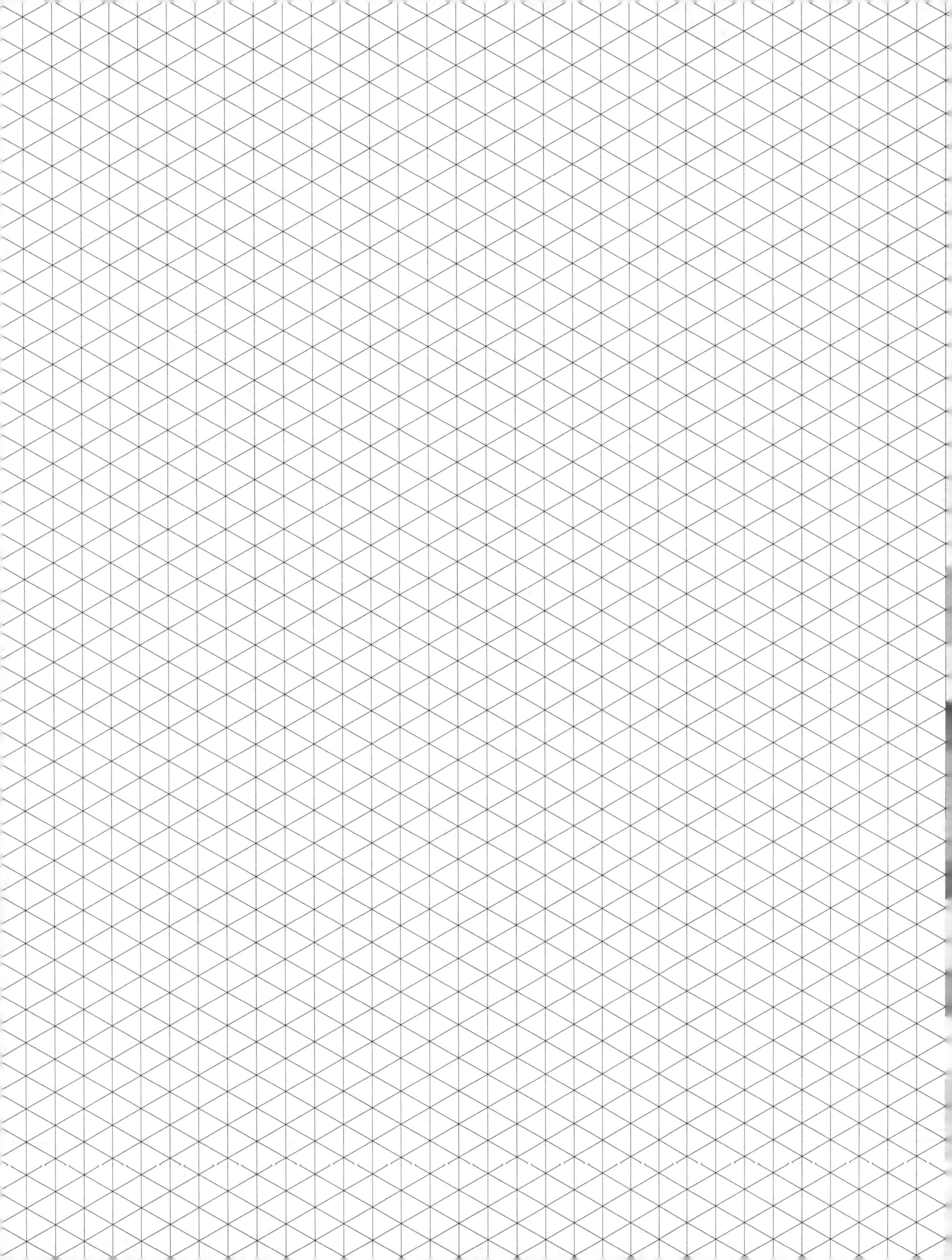

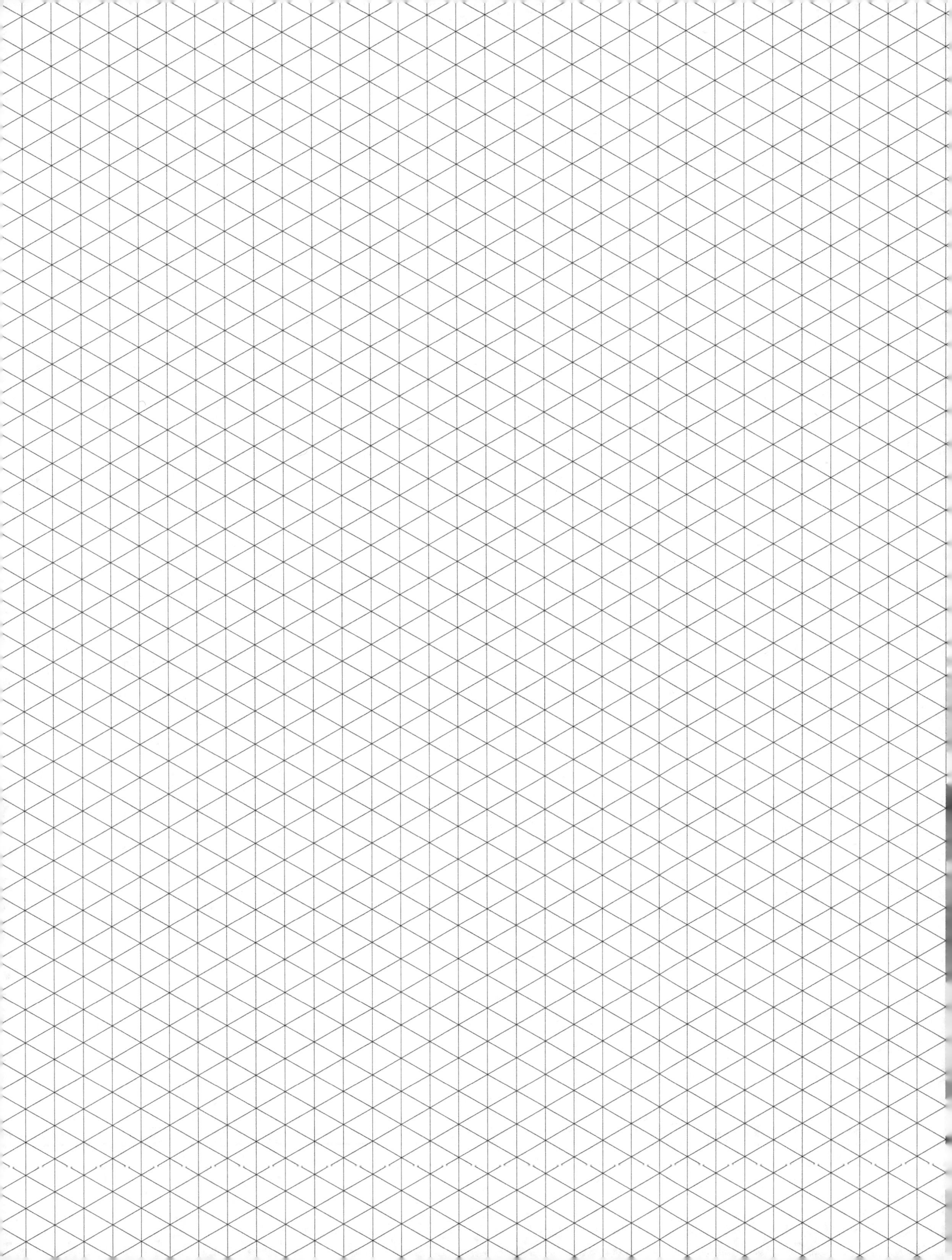

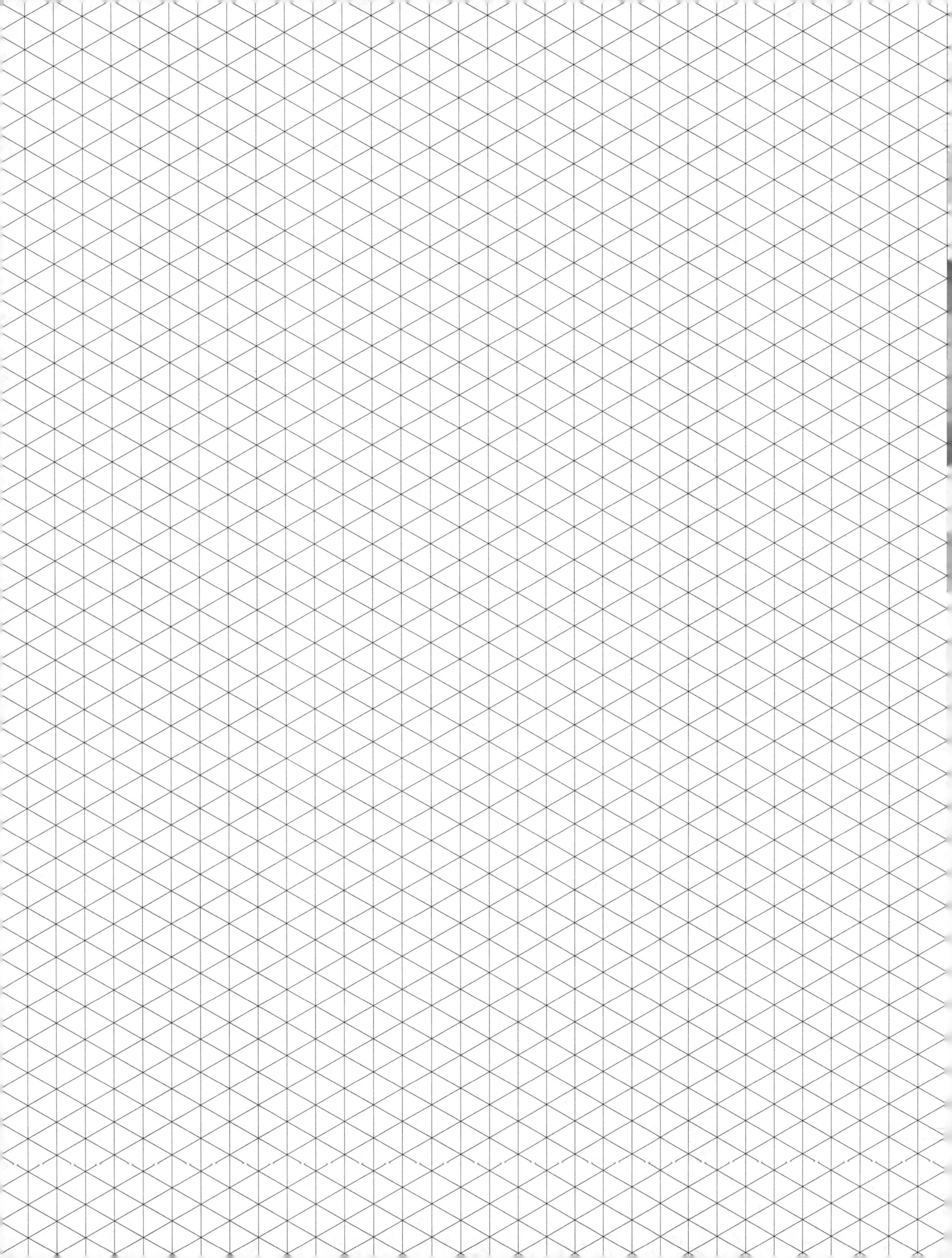

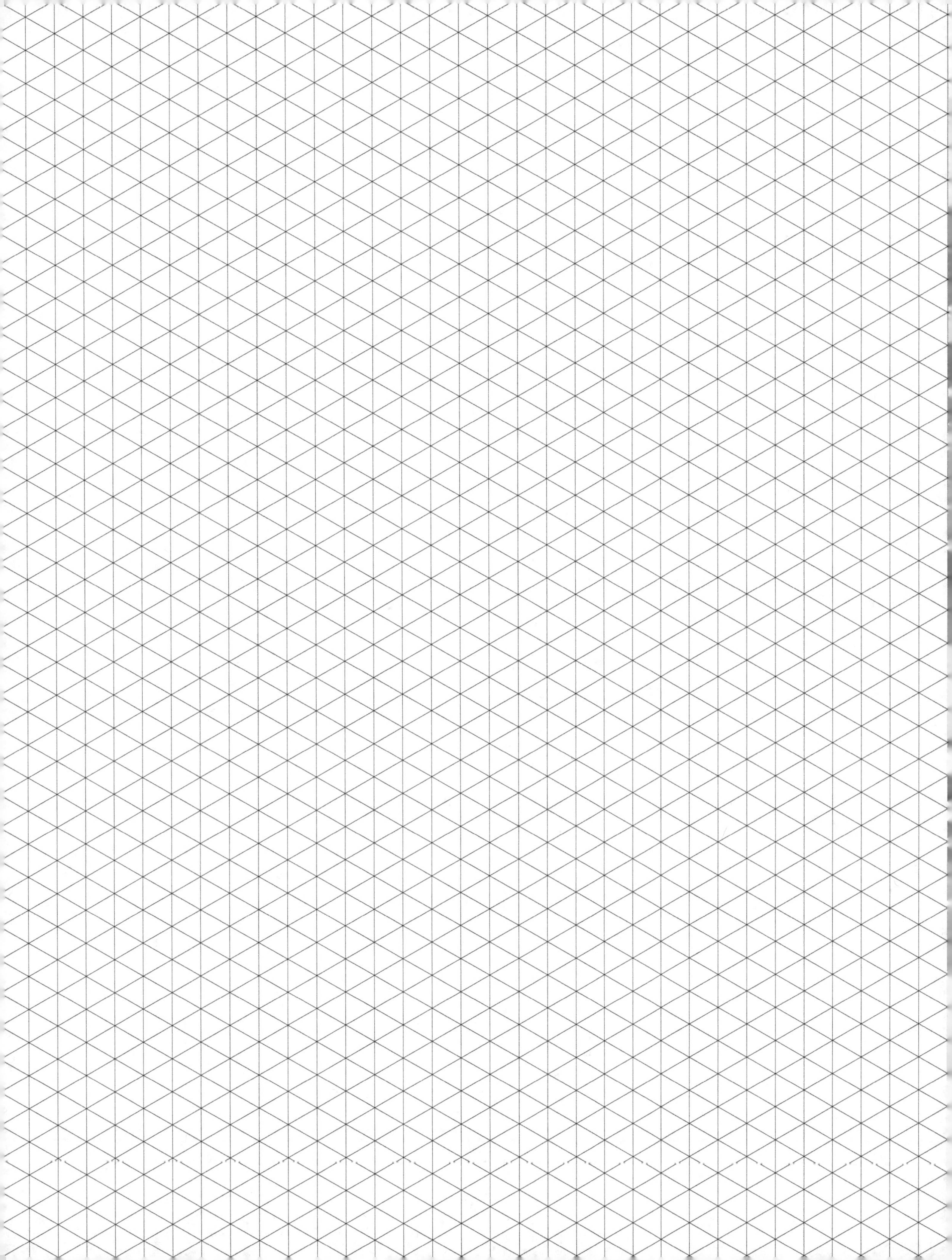

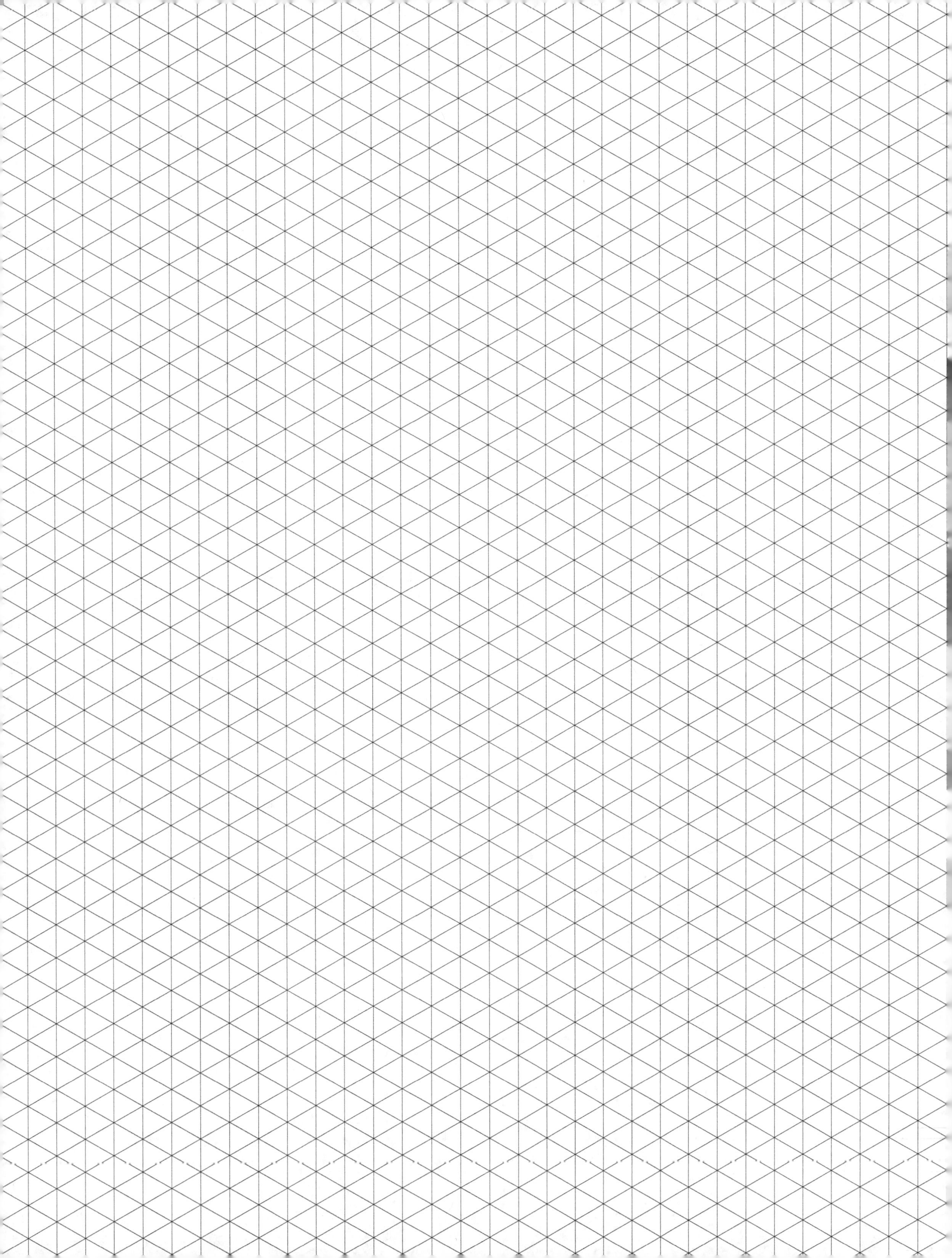

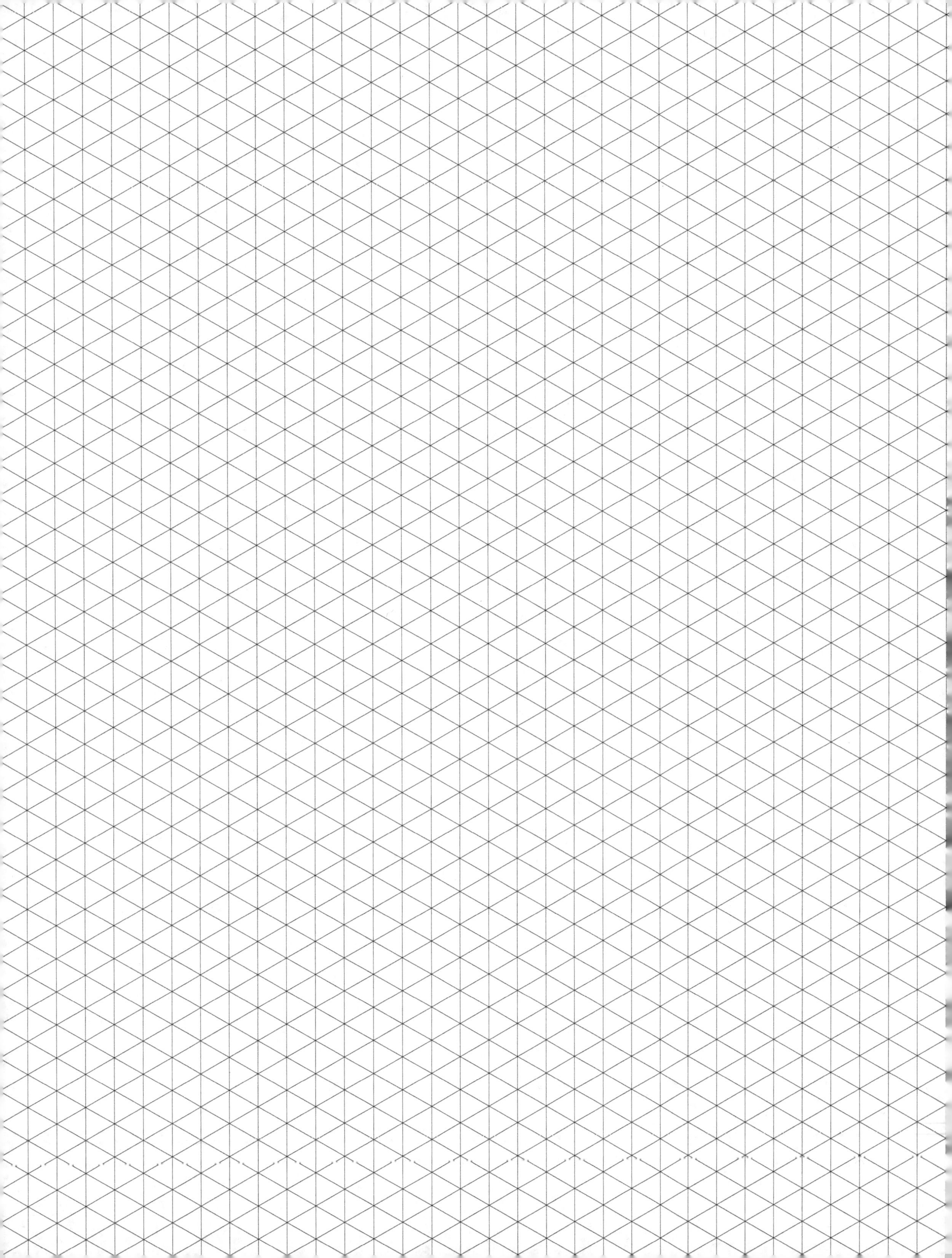

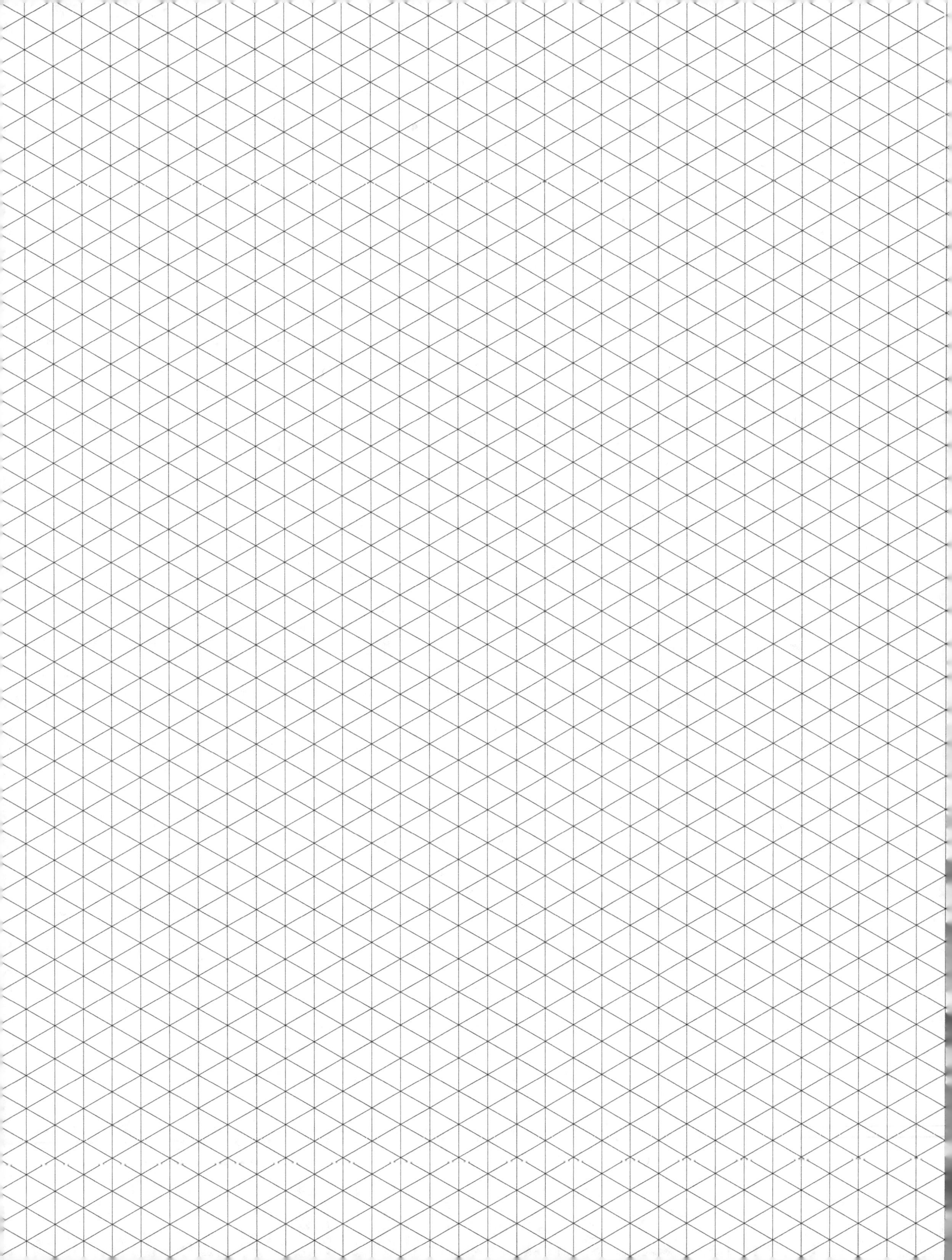

www.ingramcontent.com/pod-product-compliance
Lightning Source LLC
Chambersburg PA
CBHW082012230526
45468CB00022B/2051
9781723831003